Younoussa Moussa Petel

Management of natural resource-based IGA microprojects

Younoussa Moussa Petel

Management of natural resource-based IGA microprojects

in the western Mayo-Kebbi region of Chad

ScienciaScripts

Imprint
Any brand names and product names mentioned in this book are subject to trademark, brand or patent protection and are trademarks or registered trademarks of their respective holders. The use of brand names, product names, common names, trade names, product descriptions etc. even without a particular marking in this work is in no way to be construed to mean that such names may be regarded as unrestricted in respect of trademark and brand protection legislation and could thus be used by anyone.

Cover image: www.ingimage.com

This book is a translation from the original published under ISBN 978-620-6-71127-8.

Publisher:
Sciencia Scripts
is a trademark of
Dodo Books Indian Ocean Ltd. and OmniScriptum S.R.L publishing group

120 High Road, East Finchley, London, N2 9ED, United Kingdom
Str. Armeneasca 28/1, office 1, Chisinau MD-2012, Republic of Moldova, Europe
Printed at: see last page
ISBN: 978-620-7-65664-6

TITLE :
MANAGEMENT OF AGR MICROPROJECTS BASED ON NATURAL RESOURCES IN WESTERN MAYO-KEBBI, CHAD

DEDICATION

*To my mother **HALIMA ADAMOU***

ACKNOWLEDGEMENTS

We would like to express our sincere thanks to all those who have contributed in any way to the completion of this work. In particular, we would like to thank :

*- To **Dr FADIL MALLAYE**, for agreeing to guide this work, sharing his wealth of experience with us despite their busy schedules;*

*- To Mr. **YASSINE ASSAFO Ahmad**, National Coordinator of the RECONNECT Project, especially for the attention he paid to us during the preparation of this work;*

*- To Messrs **GABGALYAN Gaston and DIMANCHE OUIN Hervé**, respectively in charge of Local Governance and Monitoring and Evaluation in the RECONNECT project, for the information and advice they provided;*

*- To Mrs **BLAH DEBOGUE**, Secretary, for her professionalism and collaboration.*

*- To Mrs. **YAFTANE KOUMAI Juliette**, Head of the "Women Entrepreneurs" Group, for her collaboration and the data provided;*

*- To our friends **Dr MAHAMAT ABDERAHIM TOKO, Dr SEIDOU MALLOUM, Dr PASSINGRI KEUDEU, Dr ALI ABDEL RAHMANE HAGGAR, M. HAROUN MAMOUD IDRISS,** for socio-professional support;*

- To all the promoters of microprojects in the Mayo-Kebbi West Province, for their willingness to be interviewed and the quality of the information provided;

- To our parents, spouses and friends for their special interest in the production of this document;

-We would like to reiterate our sincere thanks to all of you.

SUMMARY

The "enterprising women" group in Mayo-Kebbi West has set up an income-generating activity (IGA) processing shea nuts into butter, and has obtained funding from the Global Environment Facility (GEF) through the RECONNECT project.

The management of this microproject reveals that the technical capacities of the group's members are strengthened by ongoing training for 50 people; the activity generates income for the group's members; the group's operating account is positive; the group pays 11 women and 3 young people and the schooling costs of 15 children are borne by the group.

However, there are a number of management and organisational problems. There is a lack of documents summarising activities, financial accounts need to be kept, depreciation of durable assets needs to be taken into account when calculating profits, family labour needs to be valued in monetary terms, and profit-sharing procedures need to be established. As well as the contribution to cover any operating losses, management control and the keeping of activity reports.

The study confirms the profitability of the IGA, but management capacity is limited to ensure the sustainability of the achievements at the end of the project. The strategies proposed for its improvement take account of an overall policy of consultation and ownership involving farmers in all phases of the project cycle, from design to liquidation.

Keywords : management; microproject; IGA; shea; Chad.

GENERAL INTRODUCTION

1. Background and issues

Chad's population is estimated at 17 million (RGPH 2009[1]). Some 64% of this population live below the poverty line, mainly in rural areas. GDP per capita is estimated at USD 728.34, with a Human Development Index (HDI) of 0.401. These figures place Chad among the poorest countries in the world, ranking 187th out of 189 countries (UNDP, 2019).

The development of Chad's non-renewable natural resources, in particular its oil fields, was a milestone in the country's history. It was a time rich in promises of economic growth, development and improved living conditions for the people of Chad. But the reality is mixed after a few years of exploitation.

Chad also has enormous potential in terms of renewable natural resources. However, abusive deforestation, bush fires, inappropriate farming practices, overgrazing, silting and poaching have made access to these resources difficult, resulting in competition and conflict between different users.

Despite the existence of a large hydrographic network, including 12,720 km2 of lakes, permanent rivers (Chari and Logone), semi-permanent rivers (Batha, Bahr Azoun, Salamat and Mayo-Kebbi) and numerous temporary rivers, only 18,000 ha out of a potential of over 5 million hectares have been developed.

The advanced state of degradation of natural resources and the under-exploitation of certain economic development potentials, such as the significant water resources offering a potential irrigable land area of around 5.6 million hectares, of which only 7,000 ha are irrigated, are of particular concern in the Sudanian zone (NDP, 2020).

Mayo-Kebbi West is a province of Chad whose population is predominantly rural and dependent on agriculture and livestock farming.

However, the natural environment on which these two (2) sectors depend is being adversely affected by climate change, partly as a result of human activities, including over-exploitation. The diversification of income-generating activities (IGA) for rural communities through micro-projects based on natural resources, particularly non-timber forest products (NTFPs), including shea nuts, the focus of our study, may be one of the environmental policies of the State and its partners to mitigate the scourge of degradation of the natural environment.

In this sense, support for economic micro-projects of the IGA type by programmes, projects and NGOs aims not only to increase the level of income of the rural population with a view to combating poverty, but also to encourage the users of natural resources who are the beneficiaries of this support to replace bad practices with activities that are more environmentally friendly.
"This will enable the population to contribute to the protection and sustainable management of natural resources. This will enable the population to contribute to the protection and sustainable management of natural resources.

It is undoubtedly the same philosophy that has led the RECONNECT project in Mayo-Kebbi West to provide technical and financial support for around 370 natural resource-based microprojects/AGRs initiated by grassroots organisations, including the microproject "processing shea nuts into butter" by the "enterprising women" group, which is the subject of our research.

It should also be stressed that the sustainable development of income-generating activities (IGAs) depends on their proper management and their impact on society.

However, our review of the literature shows that programmes and projects, in particular GTZ, PGRN, PRODALKA, GIZ HYDRAULIQUE, PAGL, PAFAM, etc., have recently supported community microprojects in Mayo-Kebbi West with virtually the same objectives.

But today, we have to admit that, after the departure of these organisations, very few beneficiaries continue to manage their assets and improve their living conditions.

Faced with this situation, we need to ask ourselves the following question: How is the "women entrepreneurs" group's microproject developed and managed, and what impact does it have on improving the living conditions of its beneficiaries?

2. Objectives and hypotheses

Overall, the aim of this study is to contribute to improving knowledge and facts on the management of IGA-type microprojects in West Mayo-Kebbi, in order to propose strategies likely to improve their performance.

Specifically, this involves :

- Understanding the development and management of the microproject to transform shea nuts into butter by the "enterprising women" group in Pala, Mayo-Kebbi West;
- Analyse the price of butter through marketing channels and the seasonality of supply;
- Check the profitability of the microproject through the operating account;
- Check the impact of the microproject on improving the living conditions of beneficiaries;
- Propose strategies to improve the performance of the microproject.

Three (3) working hypotheses have been put forward:

- Promoters have the technical capacity to develop and manage micro-projects;
- Marketing channels and the seasonality of supply have an impact on consumer prices;
- The microproject is profitable and has an impact on improving the living conditions of the beneficiaries.

3. Methodology

To carry out this research, we adopted a methodological approach consisting of three (3) main stages: (i) literature review; (ii) collection of empirical data; and (iii) analysis of the data. (iii) transcription, interpretation and analysis of the data:

• Firstly, for the documentary review, we conducted research in N'Djamena, at the following organisations and ministerial departments: Bibliothèque du Centre d'Étude et de Formation pour le Développement (CEFOD); Ministère de l'Environnement, de la Fisheries and Sustainable Development; Ministry of Economy and Development Planning; Ministry of Agricultural Development; Ministry of Industry and Trade; Food and Agriculture Organization of the United Nations (FAO); United Nations Development Programme (UNDP). Then in Pala to the following public and private services: Nicodème Cultural Centre Library (CCN); Provincial Delegation for the Environment; Chamber of Commerce and Crafts; Pala Municipality; RECONNECT Project; GIZ BSB-YAMOUSSA;
GIZ PRCPT; WCS; PRO-FORT; CECADEC; Association de Groupe d'Expertise Nationale en Economie Rurale (GENER); Cellule de Liaison des Associations Féminines (CELIAF) and Association des Femmes pour l'Autopromotion (AFAP).

This phase enabled us to examine several sources of information, including books, articles, journals, scientific papers, reports, archives, theses and dissertations dealing with issues related to our theme.

• Secondly, empirical data were collected through surveys carried out in three women's groups involved in processing non-timber forest products and agricultural products, namely the "women entrepreneurs" group, the "OPLO" group and the "PEE-MBANG" group. And at the weekly market in Pala, with 6 women producers, 11 merchants and 13 consumers. Interviews were then conducted with the leaders of the above-mentioned groups, RECONNECT project staff and some provincial delegates.

• Finally, these data were transcribed, interpreted and analysed using microprojects, descriptive statistics, marketing channels and the farm profit and loss account.

4. Plan

The work as a whole is structured around two (2) parts made up of four (4) chapters: Chapter 1 sets out the theoretical framework for microprojects; Chapter 2 presents the research area; Chapter 3 deals with the microproject/AGR for processing shea nuts into butter by the "enterprising women" group and, finally, Chapter 4 addresses the marketing channels, the seasonality of supply, the operating account and possible strategies for improving the performance of the AGRs.

PART ONE
THEORETICAL FRAMEWORK FOR MICROPROJECT MANAGEMENT AND PRESENTATION OF THE RESEARCH AREA

INTRODUCTION TO PART ONE

Project management is the application of knowledge, skills, tools and techniques in project activities to achieve the expectations of the parties involved in the project.
A project is a one-off, coordinated effort to achieve a single objective, including a degree of uncertainty as to whether it will be achieved (Petit Larousse).
The[2] project is the set of actions to be undertaken to meet a defined need within a set timeframe (a start and an end). The project mobilises identified resources (human and material) during its implementation, which also has a cost and is therefore budgeted for.
The microproject, like the project described by Rosanvallon[3] , can be understood as "a collective action that is purposeful, intentional and directed towards certain ends"; "a foreign aid action"; "an action that will have an impact on the economy"; "an individual action to create income-generating activities".
The "micro" scale refers to the size of the project: its reach (on a very localised scale), its impact (for a limited number of beneficiaries) and its budget (limited cost).
The microproject approach refers to the approach, the method of intervention and the philosophy of the microproject.
This first part deals with the theoretical framework of microproject management (chapter 1) and presents the research area (chapter 2).

CHAPTER I
THEORETICAL FRAMEWORK FOR MICROPROJECT MANAGEMENT

Over the last few decades, the micro-project approach has been one of the main thrusts of many development projects and programmes. Considerable human and financial investment has been made.However, it has to be said that the results have not always lived up to expectations. The response to the basic needs of communities, such as food, health and education, has sometimes been unsatisfactory or partial. Projects have often cost more and lasted longer than expected, and their effects - negative in some cases - have not always been anticipated. This can be explained in part by the fact that the activities implemented were not adapted to the socio-economic context, and by the lack of project monitoring.

SECTION I: ORIGIN AND CONCEPTS OF MICROPROJECTS PARAGRAPH I: ORIGIN OF MICROPROJECTS

The microproject originated in microfinance with Professor Muhammad YUNUS in the 1970s. This is because banks generally finance large-scale projects that not only provide them with guarantees but also generate high interest rates. Consequently, micro-projects that do not meet these conditions (guarantee, insurance, institutional framework, etc.) are excluded from the banking system.

This pertinent observation led Muhammad YUNUS to initiate microcredit, through Grameen Bank or "Bank of the Poor", in Bangladesh and paved the way for many other experiments throughout the world.

Institutions are being created to provide the poor with the means to create their livelihoods and the tools to manage the associated risk, in other words the normal financial services that are offered to wealthier categories4. The success of the Grameen Bank, which has more than 7 million poor Bangladeshis as customers, has been echoed around the world.

In practice, it has proved difficult to replicate this experience. In countries where population densities are lower, it is much more difficult to create the conditions that will make it profitable to set up local services and shops. Nevertheless Grameen Bank has shown that not only do poor people repay their loans, but they are also able to pay high interest rates, enabling the institution to cover its own costs[4] .

Perhaps that's why Muhammad YUNUS said: "I had no intention of setting up a bank, but I did wonder how the poor could improve their living conditions. In 1974, we had a famine; I was disgusted by the uselessness of the economic knowledge I was teaching. I knew someone who wanted to borrow money to develop his business, but no bank would accept. I solved the problem by lending out of my own pocket, but that was only a personal solution. I was looking for an institutional solution. I offered myself as a guarantor, got money from the bank and gave it to the people. At the same time, I came up with a few operating rules. It worked and I increased my loans from the bank. The repayment was 100%, but the bank wasn't convinced by the demonstration: what you're doing is too small-scale. It doesn't prove anything. So I did it in seven villages, but the bank didn't believe it. Then I made these loans in a whole district that the bankers had identified for me. They still weren't convinced. So I decided to set up my own bank... .[5]

It should be noted that Muhammad YUNUS was awarded the Nobel Peace Prize on 13 October 2006 for this action, which his country quickly reclaimed in 2011[6] .

PARAGRAPH II: DEFINITIONS OF MICROPROJECT CONCEPTS AND APPROACH

A- Definitions of microproject concepts[7] 1- Microproject

A micro-project can be defined as a development action initiated locally in response to the needs expressed by the beneficiaries, who are responsible for their own development.
The micro-project is of limited duration and mobilises specific resources to meet defined objectives.

The term "microproject" is used by donors for their actions in the field, as opposed to macroeconomic programmes, which are programmes ofdirect aid to the governments or major institutions of the beneficiary countries.

2- AGR microproject

An Income-Generating Activity (IGA) microproject is an economic activity involving the production and/or marketing of a good or service that generates regular income for individuals or a group of individuals (agricultural cooperative, women artisans, etc.), but also for a social structure (school, health centre, library, etc.).
It is generally designed to improve living conditions. Its implementation requires organised human, financial and material resources.

B- Micro-project approach[8]

The micro-project approach refers to the approach, the mode of intervention and the philosophy of the micro-project. It differs from the macro or programme approaches, and refers to the size of the project. The micro-project approach is very localised in scale, its impact affects a limited number of beneficiaries and its cost is limited. It is generally implemented as part of a large-scale strategic development programme.

1- Specific features of the micro-project approach

The micro-project approach prioritises specific features such as stakeholder participation, appropriate funding and rapid, sustainable results.

- **Involvement of stakeholders-beneficiaries**

Participation differs from one microproject to another, but four (4) models can be distinguished: the natural, induced, reproductive or turnkey model.

➢ **Natural model**

In this model, problems are identified by a grassroots community. The community then entrusts an organisation with setting up the micro-project. In other words, the organisation plays a supporting role in initiating the action. It cannot be the beneficiary.

➢ **Induced model**

The induced model is characterised by the fact that it is the organisation that both identifies the beneficiaries and the problems to be solved in a locality.

It is also the entity that sets up the microproject. The development process may depend on the objectives of this entity and not on the beneficiaries.

➢ **Reproductive model**

The model is reproduced when an individual or group of individuals learn about a microproject carried out in another locality. The initiators ask an organisation for help in replicating it. The relevance of the request is assessed by the organisation on the basis of the reference context.

➢ **Turnkey model**

The model involves offering turnkey solutions to local players, who will then assess whether or not they meet their needs. It is based on appropriate funding to achieve sustainable results.

2- Sectoral typology and characteristics of microprojects

- **Sectoral typology of microprojects**

Microprojects can be in different sectors of activity:

➢ Processing of agro-forestry products ;

➢ Livestock, market gardening ;

➢ Crafts ;

➢ Small businesses ;

➢ Cereal storage ;

➢ Processing of non-timber forest products; etc.

- **Characteristics of microprojects**

A microproject is characterised by the following elements:

➢ Local cooperation ;

➢ Low overall cost;

➢ A small-scale geographical impact ;

➢ A possible innovative character.

SECTION II: IMPLEMENTATION OF MICROPROJECTS[9]

Setting up a microproject such as an income-generating activity (IGA) is a guarantee of financial sustainability. Many project owners set out to create or strengthen income-generating activities to contribute to the financial viability of a social structure or to boost the incomes of disadvantaged groups. The launch process can take place in the following five (5) stages:

- Planned activities;
- Business opportunities ;
- The marketing plan ;
- The production plan ;
- Economic viability.

PARAGRAPH I: PLANNED ACTIVITIES AND BUSINESS OPPORTUNITIES

A- Planned microproject activities

The planned activities must be coherent and relevant in order to avoid social or cultural problems. They must be adapted to the area of action. To do this, before choosing an activity, it is necessary to carry out an analysis of the natural resources and the cultural and socio-economic environment. It is just as important to develop traditional activities first as it is to launch new ones.

1- Diagnosis of the microproject

The promoter must define precisely the IGA he wishes to develop and be certain that the activity, if carried out successfully, provides a relevant response to the needs expressed. The promoter analyses the real needs upstream and then considers the purpose of the action and the nature of the beneficiaries. The following preliminary questions are essential in the diagnostic phase:

- ➢ Where is the microproject located?
- ➢ What is the socio-economic context?
- ➢ What needs have been identified?
- ➢ What problems need to be resolved?
- ➢ What do you want to produce?
- ➢ What is the purpose of the IGA?
- ➢ Who will benefit from the microproject?
- ➢ Who are the stakeholders in the microproject?

- Does the proposed activity meet our needs?
- What is the added value of this activity?
- Is the activity adapted to the local context (climate, resources, culture)?
- Does the project have the support of local residents?

2- Promoting local traditional activities

It's often best to start with activities for which the local population has cultural knowledge and previous experience.

In many cases, the local population is already developing income-generating activities. The best approach for a microproject is to support these existing IGAs by helping the population to reduce the obstacles they face.

B- Commercial opportunities for the microproject

It is necessary to identify commercial outlets through market research and also to anticipate the effects.

1- Market study of the microproject

Verifying the existence of commercial outlets involves carrying out market research. The aim is to gather information in order to validate or correct hypotheses about demand, supply and the market environment. Market research involves the following actions:

- Gather sufficient information on demand and supply to enable assumptions to be made about future sales and to provide concrete elements to be used in drawing up the provisional budget;
- Set your selling price as consistently as possible;
- Choose the means of distributing your product, consider appropriate means of communication and be aware of the weaknesses of your business.

2- Forecasting the effects of the microproject

The impact that the following factors can have on the success of the microproject should be anticipated: degree of isolation and density of the road network in the area, climate, relief, seasons, availability of raw materials, presence of an electricity network, quality requirements for the selected product, available production resources, etc.

PARAGRAPH II: MARKETING AND PRODUCTION PLAN

A- Marketing plan

Defining a sales strategy means adapting the goods and services we produce to meet the needs of identified customers. In other words, it involves the decisions to be taken to achieve sales targets.

The knowledge of the market, the customer base and the competition gained from the market research enables a sales strategy to be defined.

The aim is to offer a product that satisfies its customers and to sell enough t o make a profit.

This is a delicate stage that requires thought, logic and creativity, as choices have to be made about :

• The product (image or drawing): definition of the product's characteristics and how best to meet needs (functionality, packaging, quality, etc.).
• Customers: target customers and product range (top-of-the-range, mass market, etc.).
• Distribution: choice of distribution channel and areas (place of sale, direct sale or via intermediaries, sales network, etc.). Distribution can take place through traditional commercial channels, but also through more informal, local channels or those linked to international solidarity networks.
• Communication: actions to be taken to publicise and inform consumers about the qualities and benefits of the product (advertising, promotion, sponsorship, etc.).

B- Production plan

The dimensions of the production plan take into account the commercial outlets envisaged on the one hand, and the technical and human resources available on the other:

• Construction: warehouse, shop, corridor, etc. ;

• Equipment and supplies: machinery, packaging, tools, electricity, water, etc. ;

• Raw materials ;

• Transport ;

• Workforce ;

• Training.

However, before going into production, the project owner must check the economic viability of the planned activity. They need to anticipate whether their business is likely to be profitable in the medium term, and whether it will generate more money than they spend. The economic viability of the business is measured by a projected operating account.

CHAPTER II
PRESENTATION OF THE RESEARCH AREA

The research was based in Mayo-Kebbi West on the microproject for processing shea nuts into butter run by the "women entrepreneurs" group, financed by the RECONNECT project as part of the support for microprojects and IGAs based on natural resources initiated locally by grassroots communities. In this chapter, we will provide an overview of the province of Mayo-Kebbi West and its natural resource potential, the RECONNECT project and the "women entrepreneurs" group.

SECTION I: OVERVIEW OF WEST MAYO-KEBBI AND NATURAL RESOURCES

PARAGRAPH I: MAYO-KEBBI WEST PROVINCE

A- Introduction to the province

One of the twenty-three (23) provinces of Chad, the province of Mayo-Kebbi Ouest (Decree N°415/PR/MAT/02 and 419/PR/MAT/02), whose capital is Pala, has five (5) departments, fifteen (15) sub-prefectures, twenty (20) cantons and five hundred and fifteen (515) villages covering a total area of 12,600 km2. According to INSED projections, the population of Mayo-Kebbi West is estimated at 785,944 in 2018.

The province of Mayo-Kebbi West borders Cameroon to the north and south, Mayo-Kebbi East to the north-east, Tandjilé to the east and Logone Occidental to the south-east.

It is located in the Sudanian zone between 9ème and 10ème degrees North latitude and 14ème and 15ème degrees East longitude.

B- Administrative map of the province

The figure below shows the administrative map of his constituency.

Figure 1: Administrative map of the province of Mayo-Kebbi West

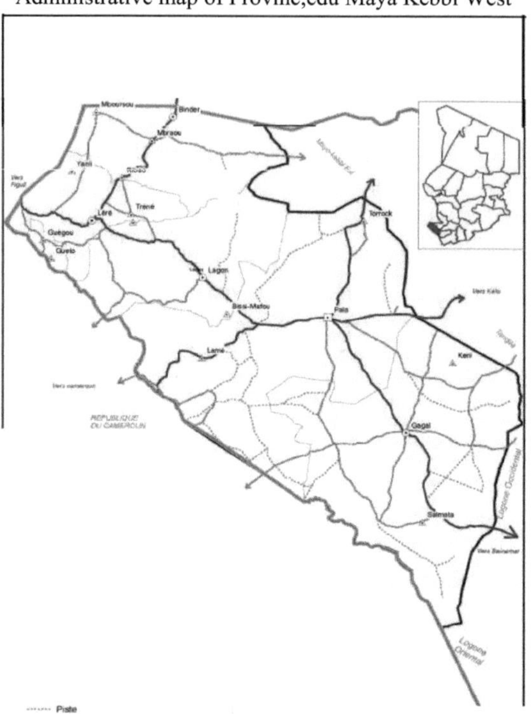

	Lege111,from Admilills T1Jllal'I [:]Gttef.l(ieu de Provil'}ce (!}Ctlef.Jlieu ded;)épartemenl CiJmmurte Fromièra d'Etat /'VLlrnle,de Pu,vlnce ; Lirntè qé Dé- i:iarternenl Camnumi tù;J,n R-oote pHnclpale nan bitumée	0 5 10 m,o:JrffliëT J. llfle-t.202	20	3ô	[)	lOnle'tèr
	--Rool.e se.coodaira	s	;rMiOHU			

PARAGRAPH II: NATURAL RESOURCES IN WESTERN MAYO-KEBBI

Defined as goods or services provided by nature without alteration by human beings, natural resources are precious to human societies. They contribute directly to their well-being and development (raw materials, minerals, food) or indirectly through the services provided by the processing of non-wood forest products.

A- Renewable natural resources

Soil, vegetation, fauna, hydrography and climate can all be cited as renewable resources.

The province of Mayo-Kebbi West has seven types of soil:
- Leached sandy-clay tropical ferruginous soils;
- Tropical ferruginous soils leached to hydro morphs at depth ;
- Hydromorphic soils on old ironclad soils ;
- Soils with a low ferral content (red soils) are very sandy and degrade very quickly;
- Tropical ferruginous soils with shallow breastworks ;
- Alluvium and hydro morphic sandy clay soils;
- Shallow or outcropping breastplates.

The vegetation consists of gallery forest, open forest, swamp forest and wooded savannah. There are three (3) protected areas in this zone: the Yamba Berte forest reserve and the two (2) National Parks of Sena-Oura and Zah-Soo. West Mayo-Kebbi is also rich in fishery resources in Lake Léré and Lake Tréné.

The fauna is particularly varied and the species found in the protected areas are representative of all the fauna in the area. These include: ourébi, gazelle, warthog, elephant, giraffe, buffalo, jackal, wild guinea fowl, ostrich, etc.

The aquatic fauna in the two lakes includes manatees, hippos, turtles, crocodiles and several other species of fish. The climate here is dry tropical with two seasons, dry and rainy.
The dry season lasts seven (7) months from November to April and the rainy season lasts five (5) months from May to October. The graph below shows the administrative map of the province of Mayo-Kebbi West.

The various natural resources in West Mayo-Kebbi include not only forest resources such as timber, but also non-timber forest products.

B- Concepts of forest products[10]

A locality's forest resources are a heritage that needs to be managed, and its potential needs to be understood. In Chad, as in West Mayo-Kebbi, non-timber forest products (NTFPs) provide emergency food during the lean season and constitute an emergency food safety net against seasonal hazards, and in cases of need for households. These products complement household food production and provide essential nutritional foodstuffs and products for medical use.

The current value of NTFPs for conservationists, foresters, development stakeholders and indigenous peoples has given rise to numerous initiatives in the area aimed at promoting the sustainable use and marketing of NTFPs as a means of improving the well-being of rural populations, while at the same time conserving forests.

• Forest

Article 13 of Law No. 14 /PR/2008 of 10 June 2008, governing forests, wildlife and fisheries resources in Chad, states that "forests are areas occupied by plant formations of trees and shrubs, excluding those resulting from agricultural activities". Timber forest products refer to products such as wood.However, the distinction between timber and non-timber forest products is not clear-cut. For practical reasons, the Ministry in charge of the environment through its Order **No.** 58/MAE/SG/DGE/2014 of 11 September 2014 created a National Advisory Committee on Non-Timber Forest Products.

• Non-timber forest products (NTFP)

Article 2 of the aforementioned law defines non-timber forest products (NTFPs) as "all forest products of plant origin other than fuelwood, service wood and industrial, user or artisanal timber".

NTFPs include leaves, fruit, seeds, nuts (including shea nuts), almonds, gums, resins, roots, bark, tubers, mushrooms, essential oils, insects, etc.

The shea nut, which is the focus of our research, is therefore a non-timber forest product and can therefore be exploited under the conditions of sustainable natural resource management.

• Shea production potential

In all producing provinces, fruit and nut collection is free of charge for off-farm stands.

There is no forestry tax on the collection or sale of shea nuts. The nuts are processed manually by women into oil and/or butter. The butter is used to make cosmetic products such as soap, ointment and cream. It is also used in the food industry.

• Production by province

The following table shows potential shea production in the seven (7) main provinces.

Table 1: Estimated potential shea production by province

No Order	Provinces	Total surface area (in ha)	Area considered (in ha)	Density (Nb/ha)	Number of feet	Percentage
1	Logone Occidental	884 542	294 847	18	5 307 251	10%
2	Logone Oriental	2 364 238	788 079	22	17 337 746	18%
3	Mandoul	1 732 664	577 555	32	18 481 741	12%
4	Mayo-Kebbi East	1 822 527	364 505	8	2 916 043	9%
5	Mayo-Kebbi West	1 283 533	256 707	12	3 080 478	10%
6	Middle Chari	4 177 754	1 392 585	26	32 207 204	29%
7	Tandjilé	1 753 624	584 541	16	9 352 661	12%
Total		4 532 313	2 866 234	-	73 079 124	100%

Source: INADES formation Tchad, BURECONS (Bureau d'Etude Conseils), 2009

This table shows that the province of Moyen Chari leads in terms of the number of shea plants (29%) and the area occupied (4,177,754 ha) by this species, followed by the province of Logone Oriental (18% and 2,364,238 ha respectively). The other provinces, including Mayo-Kebbi West, are virtually equal, with hundreds of hectares and 10% of trees in the area under consideration. The exploitation of shea nuts is based on a participatory approach to natural resources.

- **Participatory management of natural resources**[11]

Participatory natural resource management (PNRM) can be defined as a method of intervention that makes it possible to achieve sustainable management of natural resources. It is an approach that promotes the transfer of certain powers from the State to communities and other stakeholders by defining their rights, roles, responsibilities and interests. It is actually decentralised natural resource management (DNRM), the principle of which is to encourage local communities and territorial authorities to include environmental conservation and natural resource management activities in their local development programmes. The notion of participatory management of natural resources has gradually been linked to a more general methodology known as the "participatory approach".

Its main objective is to involve and closely associate local people in the diagnosis, identification, programming, implementation and monitoring of natural resource management (NRM) actions to be carried out at terroir level, and to define the responsibilities of the various partners at each stage of this process.

- **Public participation**

Involving local people means giving them back the power of initiative and decision-making in defining and implementing actions and programmes that affect their own future.

This means that external players and governments recognise farmers, livestock breeders, craftsmen, etc. as development players, partners in their own right and not as targets of an external project.

Participation occurs when a partnership or contractual relationship is established between the people affected by an action programme and the other players.

According to Gallard and Koné (1994): "it is a constantly reactivated, functional and pragmatic dynamic, in which development agents and populations combine their knowledge, know-how and will in concerted partnership actions with a view to improving, in a sustainable way, the assumption and management of the actions undertaken". The importance of the participatory approach applied to the management of natural resources is that it encourages the entire population of a village or group of villages to take effective responsibility for the restoration and development of the land.

It also ensures the establishment of a partnership for the management of natural resources at local level.

• Application of GPRN

The application of Participatory Natural Resource Management (PNRM) is a continuous process that takes into account all the phases in the life of a development project, in particular: diagnosis of the area from the point of view of forest resource management and analysis of the various constraints and priorities; planning: design and programming of the actions to be undertaken; and implementation, management and monitoring-evaluation of the entire programme.

SECTION II: OVERVIEW OF THE RECONNECT PROJECT AND THE CONSORTIUM "WOMEN ENTREPRENEURS

PARAGRAPH I: OVERVIEW OF THE RECONNECT PROJECT

Launched on 03 December 2018 in Pala for a period of 5 years, the RECONNECT project is the result of collaboration between the Republic of Chad, the Global Environment Facility (GEF) and the International Union for Conservation of Nature (IUCN). The RECONNECT project is based in Pala, the capital of the Department of Mayo-Dallah and the province of Mayo-Kebbi West. It is managed by a Project Management Unit (PMU), which is supported in the implementation of activities by the Local Project Coordination Unit (LPCU).The RECONNECT project will build on the achievements of previous initiatives and projects to adopt best practice in forestry and the management of agro-silvopastoral systems, from which West Mayo-Kebbi has benefited in the past. It will run for five (5) years. The rest of this paragraph deals with the project's objectives and structure.

The overall objective of the project is to improve the sustainable management of natural resources and forest resources in particular in order to reduce CO2 emissions and maintain ecosystem services. This objective will be achieved through :

• Improving the commitment and capacity of the various stakeholders to ensure the long-term sustainable management of natural resources, with strong involvement of grassroots communities;

• Increased CO_2 sequestration capacity through sustainable management of forest ecosystems covering 21,600 ha;
• Sustainable use of natural resources, including the development of sustainable income-generating activities and strengthening the overall resilience of communities to climate change;
• Increased production from degraded soils.

PARAGRAPH II: "WOMEN ENTREPRENEURS" GROUP

Created on 10 September 2011 and authorised to operate on 17 December of the same year, under No. 095G/MICA/CLA/10/49/2011 by the Rural Sub-Prefect of Pala, Chairman of the Local Approval Committee (CLA), the "women entrepreneurs" group is the work of a very active woman[12] with proven experience in the field of rural development and a former facilitator of the PRODALKA programme.

At present, she is often called upon as a trainer by local NGOs and projects as part of capacity-building workshops in the field of adding value to local products. These experiences of the group's founder undoubtedly contributed to the creation of this organisation and to the conditions to be met in the criteria for funding micro-projects by donors. This may confirm the hypothesis linked to experience in the trade.

The objectives of the "Femmes entreprenantes" group are as follows:
• To improve the social and economic well-being of its members;

• Create technical training sections for processing local products;

• Adding value to primary products based on natural resources, in particular non-timber forest products (NTFPs);
• Seeking outlets for finished products.

The group's area of intervention includes the Commune of Pala and the surrounding countryside, and its fundamental principle is the participation of communities with a view to empowering them in the implementation of the project, such as the identification of needs, project design and implementation, management and decision-making.

The group is made up of a number of bodies, namely the decision-making bodies and the production and management teams.

• Decision-making bodies

The General Meeting, the Board of Directors and the Executive Board are the group's decision-making bodies.

• Production and management teams

The decision-making bodies are supported by a production and management team. This team is made up of around ten people who receive regular training from the founder and partners.

The table below illustrates some of these bodies.

Table 2: Group decision-making bodies and management teams

Organs	Composition	Number of members	Tasks
Annual General Meeting	Founder and members	30	Electing the Administrative Council and Executive Board, defending the vision and the the group's mission
Board of Directors	Founder and subscribing members	5	Advisorand follow the executive
Executive committee	Founder and executive members	11	Carry out activities at group
Training team	Founder	01	Train and supervise the members
Team of production/processing	Women and girls-producing mothers	10	Transform the nuts from karities in finished products
Management team	Manager/cashier	02	Sell finished products and cash in

Source: Pala "Women Entrepreneurs" Group, 2023

The RECONNECT project team carries out activities in MKO in synergy with stakeholders, in particular the technical departments of the Ministry of the Environment, local development structures (ILOD, ADC, CVS), cantonal committees, farmers' organisations, women's and youth associations and groups (PEE-MBANG, OPLO, FEMMES ENTREPRENANTES, etc.), local communities, traditional chieftaincies and project partners (GEF, IUCN, GIZ, NOE, WCS). All these organisations are working together to improve the living conditions of the MKO population.

CONCLUSION OF PART ONE

The involvement of all social strata through meetings is crucial to the success of GPRN. This involvement is effective when all individuals contribute one hundred per cent to discussions and decision-making. Each participant must reflect and refer to his or her own experiences, knowledge and concerns in order to contribute to the group. Creativity, innovation and new ideas must prevail through honest, open and useful dialogue, leading to quality participation with maximum information.

PART TWO
PRESENTATION AND MANAGEMENT ANALYSIS OF THE "WOMEN ENTREPRENEURS" GROUP MICROPROJECT

INTRODUCTION TO PART TWO

With a view to helping people in its intervention area (Mayo-Kebbi West) to diversify their sources of income, the RECONNECT project has proposed several alternative activities in its support. Support for beneficiaries of income-generating activity (IGA) microprojects based on natural resources has been identified.

The "women entrepreneurs" group, as part of its micro-project to process shea nuts into butter, is also one of the beneficiaries.

This section addresses questions such as how were these microprojects developed by grassroots communities? And what about their management and impact on improving living conditions?

Chapter 3 presents the microproject, how it was developed and how it works, and Chapter 4 analyses the marketing channels, the operating account and possible strategies for improving performance.

CHAPTER III
MICROPROJECT FOR PROCESSING KARITE NUTS INTO BUTTER

The microproject to process shea nuts into butter is an income-generating activity (IGA) initiated by women living in the village of Zamadig, 12 km from the town of Pala. The women are members of a group known as "femmes entreprenantes". The IGA was set up to help improve the living conditions of the local population, particularly the members of the group.

The success of this microproject depends on the technical capacity of the promoters in the trade, the potential for shea species in the production area and its financing.

SECTION I: MICROPROJECTS FINANCED BY RECONNECT

Most non-governmental organisations (NGOs) or multinational development aid organisations finance and implement micro-project programmes, usually with the active participation of the host country authorities.

PARAGRAPH I: Community-scale microprojects

Micro-projects are generally community-based, touching people at the heart of their concerns and enabling them to take action to improve their situation.

The RECONNECT project has therefore opted for this "micro-project" approach in its on-the-ground support for target communities, funding 370 locally-initiated micro-projects in the 1st wave to meet the needs expressed locally by the people in the project's intervention zone.

PARAGRAPH II: Microprojects by department in the MKO

The quantitative breakdown of these microprojects by department and MKO is shown in the table and graph below.

Table 3: Breakdown of the number of microprojects by sector and department in the MKO financed by RECONNECT

No	Department Channel	El Ouaya	Lake Léré	Mayo Binder	Mayo Dallah	Nanay	Total	Costs/Micro p roject (FCFA)
1	Development of the crocodile site	1					1	5 000 000
2	Beekeeping		1	3	6	10	20	1 120 000
3	Compost	1			8		9	888 000
4	Stone cordon		29				29	735 000
5	Growing chillies			2	2		4	1 108 000
6	Fodder crops			7		4	11	1 401 500
7	Rabbit breeding				2		2	820 000
8	Small ruminant farming	9	8	3	12	19	51	820 000
9	Training				2	2	4	1 600 000
10	Improved fireplace					1	1	888 000
11	Vegetable growing	1			4		5	1 108 000
12	Mill				4		4	1 752 000
13	NTFP Pilot				1		1	3 533 000
14	Licking stone		1				1	541 000
15	Plantation	8	29	11	69	34	151	1 053 000
16	Production of plants	1	2	1	8	7	19	1 278 000
17	Drying of fish		3		2		5	1 401 500
18	Storage of cereals	1	3		8	27	39	250 000
19	NTFP processing			1	3		4	940 500
20	Forage processing						0	2 993 500
21	Knitting				3		3	888 000
22	Poultry		2			4	6	1 331 000
Total		22	78	28	134	108	370	
Percentage		6%	21%	8%	36%	29%	100%	

Source: RECONNECT project, 2022

The table shows that the number of plantation microprojects (151) is higher than all the other sectors. This is followed by small ruminant farming (51). These figures may explain the mission of the RECONNECT project, which is to improve the management of forest resources and agro-sylvo-pastoral systems through the involvement and best practice of grassroots communities. The number of microprojects per department is illustrated in the graph below.

Figure 2: Number of microprojects by Department

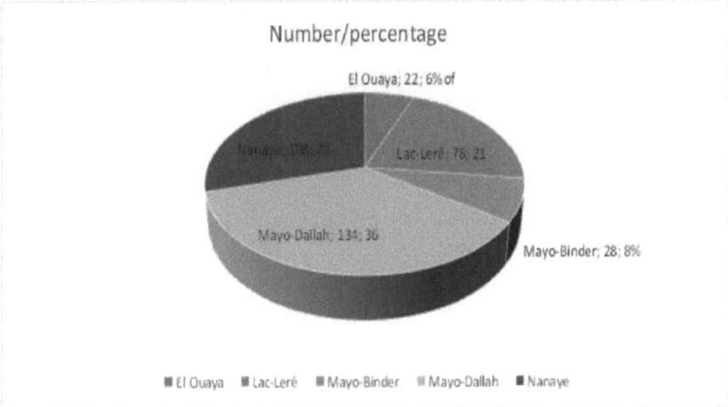

Source: **Based on data in table 3** (above)

This graph shows that the number of microprojects in the Department of Mayo-Dallah (134) comes first, followed by Nanay (108) and Mayo-Binder (78). This more or less reflects the rank of these departments, in terms of number of inhabitants, in the Project's intervention zone.

SECTION II: DEVELOPMENT OF A MICROPROJECT FOR PROCESSING KARITE NUTS INTO BUTTER BY THE "WOMEN ENTREPRENEURS" GROUP

As part of the financing of micro-projects, the RECONNECT project has provided promoters with an application form consisting of the following elements to be completed in chronological order:

- Title of the microproject ;
- Justification;
- Overall objective;
- Specific objectives ;
- Expected results ;
- Activities ;
- Schedule of activities ;
- Resources available for feasibility ;
- Simplified logical framework ;
- Budget ;
- Summary;
- Date/Name of promoter/Signatory of person responsible.

The "women entrepreneurs" group used this form to set up its microproject, the content of which is discussed in the following paragraph.

PARAGRAPH I: DEVELOPMENT OF THE MICROPROJECT FOR PROCESSING KARITE NUTS INTO BUTTER BY THE "WOMEN ENTREPRENEURS" GROUP

A- Development of the microproject

1- Justification and objectives

In order t o obtain financing for its microproject from RECONNECT, the group

"Women Entrepreneurs" has put together a dossier, the contents of which are summarised below.

- **Microproject title**

The title of the microproject is the transformation of shea nuts into butter.

- **Justification for the microproject**

In terms of natural resource management tools, women in West Mayo-Kebbi are more vulnerable than men to food shortages.

They are subject to inequalities in terms of income, education, responsibility, etc. For this reason, the abundance of shea in the locality enables women entrepreneurs to :

- ➢ Strengthen the group's technical capacity;
- ➢ Supplying butter to potential customers in terms of quantity and quality;
- ➢ Reduce the excessive cutting of trees ;
- ➢ Include women in IGAs ;
- ➢ Raising women's income levels.

- **Overall objectives of the microproject**

➢ To improve the living conditions of the women of Zamadig/Pala with t h e aim of empowering them;
➢ Contribute to improving the living conditions of the population through the self-employment of young people and young mothers in particular.

- **Specific objectives of the microproject**

Increase the income level of the Zamadig "women entrepreneurs" group by 10% in 1 year through the following actions:

- ➢ Turning shea nuts into butter;
- ➢ Selling the finished product;
- ➢ Increase the promoter's level of income ;

➢ Reduce abusive logging by women;

➢ Employ 10 young people, including 7 girl-mothers;

➢ Reducing youth unemployment;

- **Expected results**

➢ 01 fitted premises ;

➢ Acquired shea nuts ;

➢ Ingredients available;

➢ 10 tonnes of butter produced and sold;

➢ Increased income for the developer ;

➢ 10 young people and mother-girls trained and recruited;

➢ The excessive cutting of wood is reduced.

2- Activities and budget

- **Microproject activities**

➢ Fitting out the production area ;

➢ Collect and buy ingredients (shea nuts);

➢ Transporting ingredients;

➢ Provide appropriate training in microproject activities to local young people and mothers; Produce the final product (butter).

- **Activity timetables**

Month	1	2	3	4	5	6	7	8	9	10	11	12
Activities												

- **Resources available for the feasibility of the microproject**

➢ Legal recognition ;

➢ Availability of the applicant's contribution ;

➢ Shea trees in the locality ;

➢ Premises ;

➢ Existence of an open well.

- **Logical framework**

Overall objective: Improve the living conditions of women in the Erdé, Pala neighbourhood with the aim of empowering women	Specific objectives :Increase the level of Revenues of the group Erdé's "women entrepreneurs" in 1 year's time	Expected results : 1. the women of the Erdé district have come together in a "women entrepreneurs" group and their management skills are being strengthened	Activities : 1.1 Training the group in financial management
			1.2 Organise awareness-raising meetings on the benefits of the group and how it works
		2. 20 members with enhanced skills	
		3. finished products produced and sold	

- **Microproject budget**

Table 4: Budget for the microproject to process shea nuts into butter

Designation	Unit	Qté	Price per unit	Total price	Our contribution to funding		Financing request exterior
					Nature	Species	
1. Small items							
Empty 20-litre container	Unit	30	2 000	60 000	0	0	60 000
Funnel	Unit	20	500	10 000	0	0	10 000
Container	Unit	20	5 000	100 000	0	0	100 000
Plastic seal	Unit	5	2 000	10 000	0	0	10 000
All-purpose holder	Unit	5	70 000	350 000	0	0	350 000
Saucepan	Unit	10	5 000	50 000	0	0	50 000
Wheelbarrow	Unit	5	30 000	150 000			150 000
Pair of gangs	Unit	3	1 000	3 000	0	0	3 000
Subtotal 1				733 000	0	0	733 000
2. How it works							
Development at local	Unit	1	1 500 000	1 500 000	200 000	0	1 300 000
Purchase and collection of nuts	Unit	1	1 000 000	1 000 000	0	300 000	700 000
Transport from materials and nuts	Unit	1	400 000	400 000	100 000		300 000

Packaging of oil produced	Unit	1	500 000	500 000			500 000
Total operation				3 400 000	300 000	300 000	2 800 000
Unforeseen	%	0		54 200	54 200		
Grand total				4 187 200	354 200	300 000	3 533 000

Source: RECONNECT project, 2022

B- Summary of the microproject and signature of the person responsible

1- Summary of the microproject

The processing of shea nuts into butter will be carried out in the village of ZAMADIG, Erdé canton (about 10 km from the town of Pala) by the "women entrepreneurs" group, through the acquisition of small equipment and the recruitment and training of ten young people, including seven (7) girl mothers. This income-generating activity will involve collecting shea nuts in the surrounding area. These ingredients will be transported to the site where they will be processed into butter, cream and soap.

These finished products will be sold to the urban population for cash, which will increase the promoter's income and help to improve living conditions and empower young people and girl-mothers in the microproject's intervention zone.

2- Date/Name of promoter/Signature of person responsible

Pala, The Pala "Women Entrepreneurs" Group.

PARAGRAPH II: MANAGEMENT OF THE PRODUCTION AND SALE OF KARITE BUTTER BY THE "WOMEN ENTREPRENEURS" GROUP

According to the Lexique économique, an activity can be defined as an operation or process carried out by an organisation that mobilises inputs to produce outputs. Activities correspond to the direct actions undertaken or the work carried out by those carrying out the project.

Thus, the management of butter production and sales activities by group members can be the art developed by the latter to organise, direct, plan and control activities with a view to improving the group's performance strategies.

A- Knowledge of shea nuts

Before tackling the technology of shea nut processing itself, it is necessary to know about the shea tree and the uses to which its nuts, bark, leaves, wood, etc. are put.

1- Shea tree

Botanically known as Vitellaria paradoxa or Butyrospermum Parkii, the shea tree is a butter tree that grows wild in West and Central Africa, including Chad. It grows only in the Sahelian climate, with rainfall of up to 1,000 mm and two distinct seasons with a long dry period. This

10 to 15 m tall tree has a highly pivoting root system, enabling it to be used for combined crops. It takes 15 years for the tree to bear its first fruit. The edible fruit of the shea tree contains a kernel. It is from this kernel that shea butter is extracted. Shea butter is edible. The fruit is fleshy and resembles a small avocado, with a sweet, edible pulp.

It usually contains a seed surrounded by a thin shell. It takes five months to ripen and is listed as an endangered species by the IUCN. The shea tree produces an edible fruit consisting of a pulp and a nut.

2- Benefits of shea

In Chad, as in all African countries, the various parts of this tree can be exploited to produce other food, pharmaceutical, cosmetic and other products. According to Burkina Faso's national strategy for the sustainable development of the shea industry, the main types of shea products are as follows:

• The leaves are used in traditional medicine as a mouthwash or as an infusion to treat eye ailments, dental neuralgia, stomach aches and headaches. Essential oils can also be extracted. They are also used in traditional ceremonies to protect newborn babies and to make mas.

• The flowers are prepared in salads and used to make honey. They also produce essential oils.

• The bark is used in leather goods to soften skins and in traditional medicine to treat amoebiasis, leprosy and snake bites, as well as to facilitate childbirth and milk production in nursing mothers.

• Shea wood is used in the construction of houses and fences (it is highly resistant to termites) and in the production of high-quality charcoal.

• The roots are used as a remedy for diarrhoea, stomach ache and toothache.

• The fruit is prized for its very sweet pulp and fat-rich nut. The pulp is used directly for human and/or animal consumption, made into jam or used to make shea fruit juice or alcohol.

• Shea nuts contain kernels which are used in traditional medicine to combat malaria, and above all to make shea butter.

• Shea butter (conventional or organic), extracted from shea kernels, is widely used locally as edible oil for cooking and as fuel for lamps. It is also used as a raw material in the cosmetics industry, in the manufacture of a wide range of products (shampoo, face and body care, hair products, soaps, etc.).

In the food industry, it is used to make margarine or fractionated to replace cocoa butter in chocolate production. It is also used in the pharmaceutical industry. Shea also contains bioactive components that are used in a number of cosmetic products. Depending on the time of year and the region, it can also be used as fodder.

B- Processing of shea nuts by women

Shea butter comes from shea nuts, or more precisely from their fines. Shea butter is made by extracting the shea nut kernel. The kernel is washed and dried. It is then ground, refined and kneaded into a thick paste, which is mixed with water and stirred by hand.

1- Shea butter production

Butter is produced manually by women, which limits production in terms of both quality and quantity.

The head of the group tells us[13] "The women harvest the shea fruit, which contains a nut surrounded by a thin shell, containing a brown almond that contains all the fat. It is from this kernel that the shea butter is extracted by churning the paste".

This extraction method is traditional and preserves the active ingredients of the shea butter. However, the yield is lower than that obtained by chemical extraction.

2- Effects on butter quality

The factors that influence butter quality are the quality of the raw material and the method of extraction. The traditional extraction method produces butter with a bland flavour, a strong odour and high levels of moisture and acidity.

Butter made using a manual press has a less pleasant taste but a lower acidity level.

Finally, the butter obtained from the motorised press has a better colour and a more pleasant taste, and its moisture and acidity levels are very low.

The fat content is the most crucial element of the nut. If the fat content is high and the free fatty acids and moisture content are low, the exporter can receive a good price.

As far as the quality of nuts in Chad is concerned, the results of analyses of harvested kernels generally show good colouration and fatty acid content. These characteristics correspond to current standards.

The quality of shea butter depends on its intended use. The cosmetics and pharmaceutical industries prefer grade A, characterised by :

- The highest levels of skin-care substances found in the unsaponifiable or healing fraction of the butter. It should be remembered that this fraction is essentially absent from other natural oils, and this is what sets shea butter apart from other oils.
- The highest levels of the fatty acids that retain water for the skin cells and are found in the so-called saponifiable fraction.
- The chocolate industry seeks only the high levels of the four fatty acids found in the saponifiable fraction. This industry uses a chemical method of extracting the fatty acids from imported almonds. This method denatures the healing fraction but allows high fat extraction rates.

CHAPTER IV
KARITE BUTTER MARKETING CHANNELS AND PERFORMANCE IMPROVEMENT STRATEGIES

The development of a value chain depends on analysing marketing channels and improving management and decision-making performance. This chapter examines the different marketing channels, from producer to distributor to consumer, and their impact on shea butter prices. It also analyses the seasonality of supply, the socio-economic effects through the operating account before proposing strategies to improve the performance of the activities.

SECTION I: MARKETING CHANNELS FOR KARITE BUTTER AND IMPACT ON THE CONSUMER PRICE

The marketing of shea butter may involve several economic agents, namely the producer, the trader, the exporter and the consumer. The categorisation of circuits depends on the number of agents of which they are composed. These channels have an impact on the consumer price.

PARAGRAPH I: MARKETING CHANNELS

A marketing channel is a network linking producers to consumers, either directly or via other economic agents.

A- Different circuits

We distinguish four types of channel in the case of the marketing of shea butter by the Pala group. The economic agents involved in these different channels are the producer (the Pala 'enterprising women' group), the merchants and the consumers. The following table illustrates the percentage of supply according to the different channels.

Table 5: Monthly supply by marketing channel

Purchaser	Consumer (Circuit 1)	Merchant local (Circuit 2)	Local merchant (Circuit 3)	Merchant Exporter (Circuit 4)	
Place of exchange	On site (Zamadig)	On site (Zamadig)	At the weekly market in Pala (Badadji)	At the weekly market in Pala (Badadji)	Total
Monthly supply (in litre)	10	20	65	5	100
Percentage	10%	20%	65%	5%	100%

Source: Pala weekly market survey, 2022

• **Circuit 1**

This is a circuit in which the transaction takes place directly between the producer and the consumer.

Producer ⟶ Consumer

Occasionally, a private individual will go to the group's site to buy butter for household consumption. This type of exchange is often carried out by women for their food preparation. Channel 1 carries around 10% of the supply.

• **Circuit 2**

It's a circuit that links the producer to the merchant.

Producer ⟶ Merchant

This is where the merchant comes to the production site for supplies. This channel is widely used by merchants who have the privilege of meeting the producer easily. Channel 2 carries around 20% of the supply.

• **Circuit 3**

Producer ⟶ Retailer ⟶ Consumer

Channel 3 includes the producer, the merchant and the consumer.

In this circuit, the merchants, who are traders, obtain their supplies from the producer who has butter on the weekly market.

They then sell to consumers elsewhere, or sometimes on the same market. Channel 3 carries around 65% of the supply.

• **Circuit 4**

Unlike channel 3, channel 4 substitutes consumers for exporters, who purchase products from merchants on the market.

Producer ⟶ Merchant ⟶ Exporter

In this circuit, instead of the merchant delivering his products directly to the consumer, he delivers them to the foreign exporter. It is the exporter who resells the products outside the country, particularly in Nigeria. Channel 4 carries 5% of supply. It should be noted that circuit 3 carries the largest proportion of supply (65%). This is a typical feature of food markets in Chad, particularly in Mayo-Kebbi West. The presence of intermediaries is very noticeable in these markets, and consequently contributes to the rise in the price to the buyer. The figure below illustrates the different circuits involved.

Figure 3: Shea butter distribution channels for Pala producers

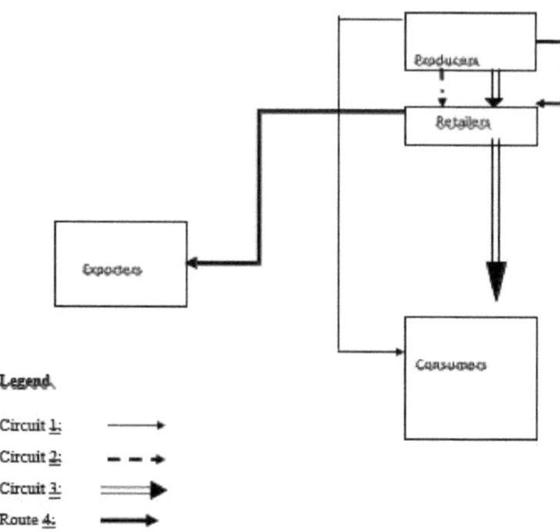

Source: Shea butter market survey of Pala producers, 2022

B- Impact of the channel on the consumer price

The consumer price varies from one channel to another. This variation is due to marketing costs. The table below illustrates this point.

Table 6: Average price of a litre of shea butter (in CFA francs) by channel

Circuits	Prices at producer	Number intermediaries	Consumer prices	Differential of Price
1	1 700	0	2 000	300
3	2 750	1	3 500	750

Source: Shea butter market survey of Pala groups, 2022

Producer prices for channels 1 and 3 are CFAF 1,700 and CFAF 2,750 respectively. The consumer prices for these same circuits 1 and 3 are CFAF 2,000 and CFAF 3,500 respectively. Through the various channels, the consumer price is set taking into account marketing costs.The merchant incurs marketing costs, which is why the price differential for channel 3 is higher than for channel 1. Channel 1, on the other hand, has a low price differential (300 FCA). This can be explained by the low marketing costs incurred by the producer in this channel.In addition, the consumer does not appear in circuits 2 and 4, which is why they have not been discussed here.

PARAGRAPH II: SEASONALITY OF SUPPLY AND OPERATING ACCOUNT

A- Seasonality of supply

The supply of shea butter on the Pala market is variable from one season to the next. The following table shows the supply statistics for the period 2022-2023.

Table 7: Supply statistics for the 2022-2023 period

Quarterly seasonality	April-June 2022	July-Sept. 2022	October-Dec.2022	January-March 2023	Total
Production (in litres)	875	945	910	770	100
Percentage	25%	27%	26%	22%	100%

Source: Shea butter market survey of Pala producers, 2023

This table shows that the differences between these different quarters are not significant. However, we can distinguish three supply periods in the year: the recovery period (April-June); the period of abundance (July-September) and the period of scarcity (October-December and January-March). These periods are illustrated in the graph below.

Figure 4: Seasonality of supply

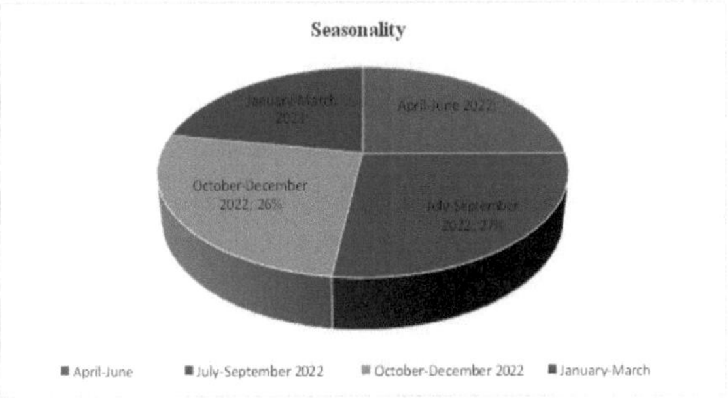

Source: From table 7 (below)

- **Period of abundance**

The period of abundance is from July to September. This is the period when the shea tree produces enough fruit. This means that enough shea nuts can be harvested for processing.

As a result, there is an abundant supply of finished shea products, particularly butter, on the market.The figure below shows the percentages of quantities sold through various channels during the period of abundance.

Figure 5: Percentages of quantities sold through various channels during the period of abundance.

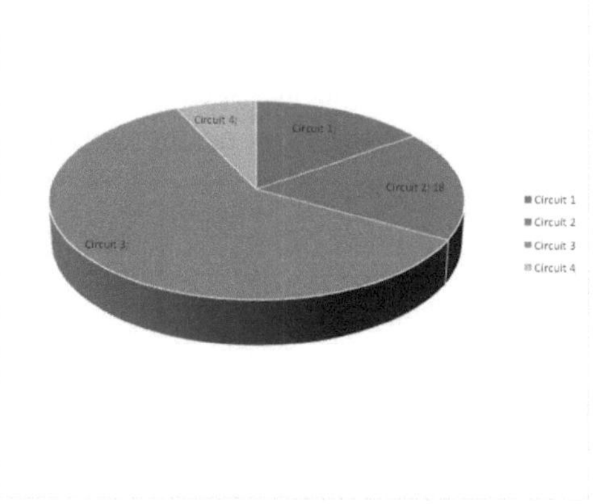

Source: From Table 7 (below)

- **Period of rarity**

The scarcity period is when the supply of butter on the market is very limited. It corresponds to the months of October to February and August. Processing activity is based solely on nuts stored during the collection period. This causes the price of NTFPs to soar, and consequently the price of shea butter in Pala. The figure below shows the quantities sold during the scarcity period through various channels.

Figure 6: Percentages of quantities sold through various channels during the scarcity period

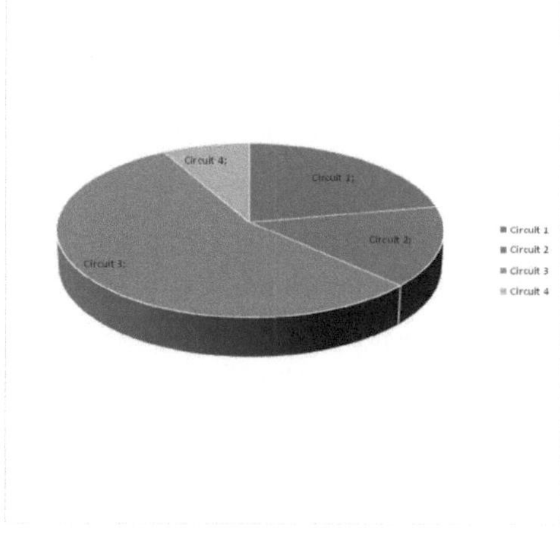

Source: From Table 7 (below)

- **Recovery period**

The recovery period is the period during which the quantity of shea butter on the market gradually increases until it reaches the period of abundance. It covers the months of April, May and June. The figure below illustrates the quantities sold by channel during this period.

The recovery period is when the rains start to fall in the production areas. This allows the plants to flower.

Figure 7: Percentages of quantities sold on various circuits during the takeover period

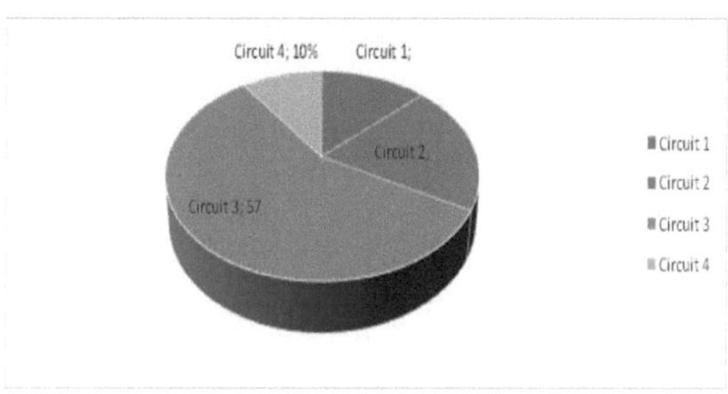

Source: From Table 7 (below)

The different periods of supply analysed above are not irreversible: a period of scarcity can be reversed by a natural phenomenon such as rainfall in a supply area. On the other hand, a social crisis can overturn a period of abundance. Each of the above periods provides income to the promoter according to an appropriate farm account.

B- Tour statistics by seasonality

Table 8: Circuit statistics by seasonality

Period	Circuit 1	Circuit 2	Circuit 3	Circuit 4	Total
Abondance	15%	18%	60%	7%	100%
Rarity	22%	15%	55%	8%	100%
Takeover	13%	20%	57%	10%	100%

Source: Shea butter market survey of Pala producers, 2023

This table shows that channel 3, in which the merchant is the intermediary between the producer and the consumer, dominates in all seasons.

SECTION II: PERFORMANCE IMPROVEMENT STRATEGY PARAGRAPH I: OPERATING ACCOUNT AND SOCIO-ECONOMIC IMPACT

A- Operating account

An operating account is the table that summarises expenditure or costs (what you spend) and income or revenue (what you take in after the sale). It is used to calculate the result, which is the difference between income and expenditure. It is also used to assess profitability, which is the ratio between profit and cost or sale. The following table shows the profit and loss account for 100 kg of shea nuts sold by the "women entrepreneurs" marketing group.

Operating account for 100 kg of shea nuts

Expenses					Products				
Designation	Unit	Qty	P. U	P. T	Designation	Unit	Qty	P. U	P. T
Buy a bag of	Unit	1	2	2	Sale of	Litre	35	2	70
Nuts from			000	000	Butter from			000	000
100Kg					shea				
Buy can	Unit	10	100	1					
Water from 20				000					
litres									
Buy	Unit	35	50	1					
packaging				750					
(container of									
1litre)									
Amortisatio n	Unit	1	26	26					
nt and Expenses			330	330					
fixed assets14									
Main	Unit	3	2	6					
d'Œuvre			000	000					
daily									
3 people.									
Profit	32 960								
Total	70 000				Total	70 000			

Commercial profitability: Profit margin/Total sales= 0.47 i.e. 47%.

Economic profitability: Profit margin/Total cost= 0.89 i.e. 89%.

- Total cost: 37,080 CFA francs;
- Group profit margin per capital invested: 0.88, i.e. 88%.
- Profit margin per day worked by the group: 32,920 CFA francs.

For 100kg of shea nuts processed into butter, the producer (group) generates a profit margin of 32,920 CFA francs against 37,080 CFA francs invested. This means that he earns 0.88 CFA francs for 1 CFA franc invested per working day. In other words, for every 100 CFA francs spent, he earns 88 CFA francs per working day. These figures confirm the hypothesis that the IGA of processing shea nuts into butter is profitable.

B- Socio-economic impact

In terms of the income received from their micro-project, which was set up to help the women in the group to earn an income, it appears that some of them have seen their incomes increase, particularly those who have actively participated in the production of finished products.

Against a backdrop of widespread poverty, which forces husbands to spend long periods away from their families in the hope of finding a better life elsewhere, women's income can play a number of roles.

Firstly, in terms of taking responsibility for certain items of expenditure that might otherwise have fallen to the man alone. The contribution of women makes life easier, particularly in terms of ensuring that emergencies are taken care of within the family. They will be able to take care of the household's food needs without having to wait for the husbands' contribution. They will also contribute to the payment of school fees or the health of their children.

The gains available to them have certainly helped to make them more confident and self-assured. We can see a kind of regression in the domination of husbands over wives, due to their new status as economic providers.

The fact that women have an income has an impact on improving their status and living conditions. This results in a less unbalanced power relationship with their husbands. The women in the "women entrepreneurs" group acknowledged that men and society are looking up to them more and more, as soon as they are able to take certain responsibilities in hand. But it remains clear that, even with relative economic autonomy, the burden of women's domestic work remains intact.

PARAGRAPH II: MARKETING TOOLS AND PROPOSED STRATEGIES

A- Marketing tools

According to Porter[15] , entrepreneurs generally use certain tools, such as product, price, point of sale and promotion, in their strategies to improve the performance of their business. This strategy is known as the "4 Ps policy".

- **Product**

This involves selecting products and presenting their quality to customers.

- **Prices**

Define the price of the product, taking into account all the elements that go to make up the price, such as production, processing and marketing costs.

- **Place of sale**

The place of sale and the size of the market need to be known to boost your sales further.

- **Promotion**

Promoting means making customers aware of your products in order to entice them to buy.

Commercial strategy

According to Marine Lalique[16] , the definition of a commercial strategy refers to the adaptation of the goods and/or services to be produced in order to satisfy the needs of identified customers. In other words, it involves the decisions to be taken to achieve sales targets.

The knowledge of the market, the customer base and the competition gained from the market research enables a sales strategy to be defined.
The aim is to offer a product that satisfies customers and to sell enough to make a profit. It's a delicate stage that requires thought, logic and creativity, because choices have to be made about :

- The product (image or drawing): definition of the product's characteristics and how best to meet needs (functionality, packaging, quality, etc.);
- Clientele: target clientele and product range (top-of-the-range, mass market, etc.) ;
- Distribution: choice of distribution channel and areas (place of sale, direct sale or via intermediaries, sales network, etc.). Distribution can take place through traditional commercial channels, but also through more informal, local channels or those linked to international solidarity networks;
- Communication: actions to be taken to publicise and inform consumers about the qualities and benefits of the product (advertising, promotion, sponsorship, etc.).

B- Proposed strategies to be undertaken

Strategies to improve the performance of IGAs with a view to their sustainability necessarily involve analysing problems and their causes. However, they must be relevant, coherent, efficient, feasible and produce sustainable results. The summary is presented in the table below.

Summary of strategies for improving the performance of an IGA

Problems	Causes	Strategies to be undertaken
Limited information to meet market needs	Discretion on the part of operators, most of whom are women, in terms of information and communication	Promote the farmers' culture by organising festivals such as dances, archery, horse races, etc.
Limited involvement of operators in the sales process	Influence major from intermediaries	Professionalise intermediaries within a formal framework
Women farmers not given much consideration in trade	Prestige, subsistence farming: products for family consumption, ...	Continuously raise awareness of the campaign fund among women farmers to convince them of the benefits of exchanges
Many projects but less efficiency	Unsuitability of project objectives and decision-making policy	Involve operators in all phases of the project cycle, from design to winding-up, at expert conferences, etc.
Le FR does not sufficiently cover WCR	Seasonality has risen sharply	Increase the funds associative

Shea butter marketing activities operate through four (4) main channels. Of these, channel (3), which links three (3) economic agents, namely the producer, the trader and the consumer, carries most of the supply (65%). The consumer price in this channel is the highest (3,500 FCFA/L) compared to the other channels, explained by the existence of intermediaries. However, the period of abundant supply contributes to the drop in price. The IGA has created several jobs, and its operating account shows a substantial profit margin (32,950 FCFA/working day). This helps to improve the living conditions of the group's members.

GENERAL CONCLUSION

At the end of our research on the "enterprising women" group, through its microproject (or IGA) for processing shea nuts into butter, the following points emerged:

- The development of the microproject starts with the identification and justification of problems, through the definition of objectives, expected results, activities and timetable, to cost budgeting.
- The technical capabilities of the group's members are strengthened through ongoing training (50 women).
- The variation in the price of butter depends on the marketing channel and the seasonality of supply (the price varies between 2,000 and 3,500 FCFA).
- The business of processing shea nuts into butter is profitable (88%).
- The group's operating result (for 100 kg of shea nuts) is positive (32,760 FCFA).
- The group pays its members, especially those who take an active part in production work (11 women and 3 young people).
- The group contributes to the cost of schooling (15 children).

However, there are a number of management and organisational problems. There is a lack of documents summarising activities, financial accounts are not kept, depreciation of durable assets is not taken into account when calculating profits, family labour is not valued in monetary terms and there are no procedures for sharing profits.As well as the contribution to cover any operating losses, management control and activity reporting. There is also confusion between turnover, profit and profitability; between IGA results and membership fees; and underestimation of the preparation time required for production. From all of the above, we can say that 3 out of 4 hypotheses (i.e. 75%) are confirmed. These are the impact of the circuit and seasonality of supply on the price of butter; the profitability of the microproject and its impact on improving the living conditions of the beneficiaries. However, the realities on the ground show us that the promoter's microproject management skills are weak in order to successfully run and sustain the IGA after the donors have left. The strategies proposed to improve the performance of IGAs, in particular the professional and socio-cultural promotion of producers and the strengthening of their capacities and permanent capital, can only be effective through a comprehensive policy of consultation and ownership involving farmers in all phases of the project cycle, from design to winding up. However, there are still many challenges to be met when it comes to environmental issues and sustainable development. So is Research and Development in this area still relevant?

BIBLIOGRAPHY

Dr TEBANI, 2015. Course in Project Management, University of Hassiba Ben Bouaali-Chlef.
KHANDKER, SHAHIDUR R. Fighting Poverty with Microcredit, - Bangladesh edition, The University Press Ltd, Dhaka, 1999.
BOYE, Sébastien - HAJDENBERG, Jeremy - POURSAT, Christine. Le guide de la microfinance, Eyrolles, 2006.
FOURNIS Y., 1974. Les études de marché : Techniques d'enquête, de questionnaire, de sondage et de contrôle des résultats Marketing, 2nd edition- Bordas, Paris

MARINE L., Comment mettre en œuvre une activité génératrice de revenus dans le cadre d'un microprojet, édition-GER, 2016

Dr MALO D., 1999. Cours de microéconomie, University of Bangui.
Pr MBETID-BESSANE E., 2004. Rural Econometrics Version I, University of Bangui.
MOURLHON M. and POULTEAU S., 1997. Management et Gestion commerciale 1ère et 2ème année. BTS force de vente, published by Hachette-Livre, Paris, March 1997.

PETEL Y, 2019. Formation du prix du bétail au marché de Pk 13, Bangui, EUE Projet
RECONNECT, " Manuel des procédures pour la mise en place du système de suivi-evaluation", 2018

RECONNECT project, "Communication strategy 2020-2023", February 2020
RECONNECT project, "Guide d'animation et d'appui pour la promotion de meilleures pratiques agricoles", August 2019

RECONNECT Project, "Report of the study for the implementation of the Resilience and Adaptation and Assessment of Transformation (RAPTA) pathway framework", October 2019
FAO, "Manuel du technicien, Gestion participative des ressources naturelles : démarches et outils de mise en œuvre", 2004

GTZ, "Auto promotion dans la gestion des ressources naturelles. Rapport de l'atelier de réflexion et de concertation tenu à Natitingou", 26-28 May 1999.

Agence Micro Projet "Designing and setting up an international solidarity microproject" Paris, January 2016
HAMADOU M et YAYE M, 1994, Analyse des méthodes et outils d'intervention des projets en milieu rural WEBOGRAPHIE
www.agencemicroprojets.org
http://www.interaide.org/pratiques/

https://fr.wikipedia.org/wiki/Beurre_de_karité

APPENDIX

APPENDIX 1: QUESTIONNAIRES FOR MICROPROJECT PROMOTERS PROJECTS

Full name:
Title: Contact:
Objectives and players
What is the RECONNECT Project? What are the objectives of the project?
What are the components of the project? Can you describe the tasks of each component?
Participation in the project

How was public participation taken into account during the subsequent stages of the RECONNECT project?

Identification Preparation
Ex ante evaluation Implementation
What difficulties did you encounter during these stages?
Development of micro-projects
What are the steps involved in preparing a microproject? What conditions need to be met to benefit from a microproject based on natural resources, in particular the processing of shea nuts into butter?

Capacity building for promoters
What are the training topics and target audience?
What techniques do you use to enable promoters to monitor and evaluate microprojects?
Monitoring microprojects

How are IGA-type microprojects monitored?

Are microprojects monitored by the technical departments? If YES, how?

Making developers accountable
What measures have you put in place to enable promoters to continue operating?

Difficulties encountered and suggestions
What difficulties have you encountered? And what suggestions do you have? Thank you for your help!

APPENDIX 2: QUESTIONNAIRE FOR MICROPROJECT PROMOTERS

Full name:
Title:
Contact :
How did you hear about the RECONNECT Project? What motivated you to set up your microproject? What is the overall cost of your micro-project? How much did you contribute? What do you think of this amount?
Are you already working in the sector before you can benefit from RECONNECT funding?

Are you a member of a group?

How do you feel about cohesion within the group?
Based on your experience, are you ready to invite other people to set up microprojects?

? Justify
Have you benefited from RECONNECT training?
Are you satisfied with the training you have received?
Would you have liked other themes?
Does your microproject enable you to meet your needs? If so, which ones?

Are you satisfied with RECONNECT's procedures in terms of :
-Selection of microprojects ?
Financing micro-projects?
How do you manage to control your activity?
Are you experiencing any difficulties in implementing your microproject? Do you have any suggestions for resolving these difficulties?

Specific questionnaire for the promoter of the microproject for processing shea nuts into butter

What motivated you to set up your IGA?
Why did you create it in Pala and not elsewhere?
What are your technical skills in processing shea nuts into butter?

Is this business profitable?
If so, how much do you expect to earn at the end of an operating cycle and then per month?
What are your short-term development prospects?
Will your IGA be able to develop rapidly?

Thank you for your cooperation

Table of equipment and sustainable expenses for the women entrepreneurs group

Designation	Unit	Quantity	Unit cost	total cost	Service life	Amorti. or load/year	real cost	cycle	Amortization/cycle
Premises costs	Unit	1	1 300 000		20	65000,00	65 000		
All-purpose holder	Unit	1	70 000		10	7000,00	7 000		
Wheelbarrow	Unit	1	35 000		10	3500,00	3 500		
Plastic seal	Unit	1	2 000		3	666,67	667		
Funnel	Unit	1	500		3	166,67	167		
Saucepan	Unit	1	6 500		3	2166,67	2 167		
Pair of gangs	Unit	3	3 000		3	1000,00	1 000		
							79 500	3	26 500

Source: Based on surveys of shea butter producers

This table shows the calculation of equipment depreciation or lasting costs in the production of butter from 100 kg of shea nuts.

APPENDIX 3: SIMPLIFIED SHEET FOR SETTING UP MICROPROJECTS/RECONNECT

Restoration of ecological corridors in West Mayo-Kebbi in Chad, in support of multiple land and forest benefits (RECONNECT) project

	Date :
Reference number (this part is reserved for the jury)	Title of the microproject :
Province Department Township / Village	
Name and legal status of the promoter (OP, GIE, ACD, ILOD ...) :	
Promoter's address (place of action):	
N° (Part reserved exclusively for the jury) :	
Developer's experience:	
Membership of a professional association: (if so, which one(s) (union, federation, platform, etc.)	
Brief description of the microproject: Location of the microproject: Justification: Objectives : Overall: Specific: ... Target sector or type of IGA: Microproject activities: Expected results: Microproject management structure: Estimated cost of the microproject (including the beneficiary's contribution in cash and in kind):Total cost of the project in CFA francs: Applicant's contribution in CFA francs: Grant requested in CFA francs: Implementation period in months: Requested start date in DD/MM/YY:	
Partnerships envisaged by the microproject:	

I, the undersigned, Mr/Mrs certify on my honour that the above information is accurate.

NB: All applications must be submitted to the prefectures in the departmental capitals. No applications will be received by the project. For further information or guidance, please contact.

The Promoter

(Full name, date and signature)

APPENDIX 4: MICROPROJECT/RECONNECT FUNDING APPLICATION FORM

Date: //

Name of promoter: //
Full address: //

A RECONNECT Pala Coordinator
Subject: Funding application
Mr. Coordinator,

We have the honour of requesting funding for the implementation of the microproject (indicate the title of the microproject)
The estimated cost of which is CFAF
(Indicate the total amount, including equipment or infrastructure, operating and other recurring costs, etc.).
As part of our contribution, we undertake to provide, from our own funds[17] , the sum of (indicate the amount of the promoter's financial contribution) : For the implementation of this microproject.

This amount will be deposited in account number (indicate account number) : at (give name and address of institution)

An actual contribution in kind is available in the form of work, services, etc. for an estimated value of FCFA Attested by the attached certified form.
This contribution, the fruit of our own efforts, represents .% of the total cost of our microproject.

The Project Sponsor (Name and signature)

TABLE OF CONTENTS

Printed by Books on Demand GmbH, Norderstedt / Germany